ANIMALS OF THE GRASSLANDS

GIANT GIRAFFES

THERESA EMMINIZER

PowerKiDS press

New York

Published in 2021 by The Rosen Publishing Group, Inc.
29 East 21st Street, New York, NY 10010

Portions of this book were originally authored by Mary Molly Shea and published as *Giraffes*. All new material in this edition authored by Theresa Emminizer.

Editor: Theresa Emminizer
Book Design: Michael Flynn

Photo Credits: Cover, p. 1 ArCaLu/Shutterstock.com; series background Chantal de Bruijne/Shutterstock.com; p. 5 StanislavBeloglazov/Shutterstock.com; p. 7 (feet) Bokeh Blur Background; p. 7 (head) Oleg Kozlov/Shutterstock.com; p. 9 Mikko Hyvärinen/Shutterstock.com; p. 11 Craig Fraser/Shutterstock.com; p. 13 Pisit Rapitpunt/Shutterstock.com; p. 15 Moonborne/Shutterstock.com; p. 17 Wim Hoek/Shutterstock.com; p. 19 EcoPrint/Shutterstock.com; p. 21 Volodymyr Burdiak/Shutterstock.com; p. 22 jaroslava V/Shutterstock.com.

Cataloging-in-Publication Data

Names: Emminizer, Theresa.
Title: Giant giraffes / Theresa Emminizer.
Description: New York : PowerKids Press, 2021. | Series: Animals of the grasslands | Includes glossary and index.
Identifiers: ISBN 9781725316980 (pbk.) | ISBN 9781725317000 (library bound) | ISBN 9781725316997 (6 pack)
Subjects: LCSH: Giraffe–Juvenile literature.
Classification: LCC QL737.U56 E44 2021 | DDC 599.638–dc23

Manufactured in the United States of America

CPSIA Compliance Information: Batch #CSPK20. For Further Information contact Rosen Publishing, New York, New York at 1-800-237-9932.

CONTENTS

Tallest of All

Giraffes are the tallest land animals in the world. They can grow up to 19 feet (6 m) tall. That's taller than three adult people standing on top of each other! Giraffes live in the **grasslands** of Africa.

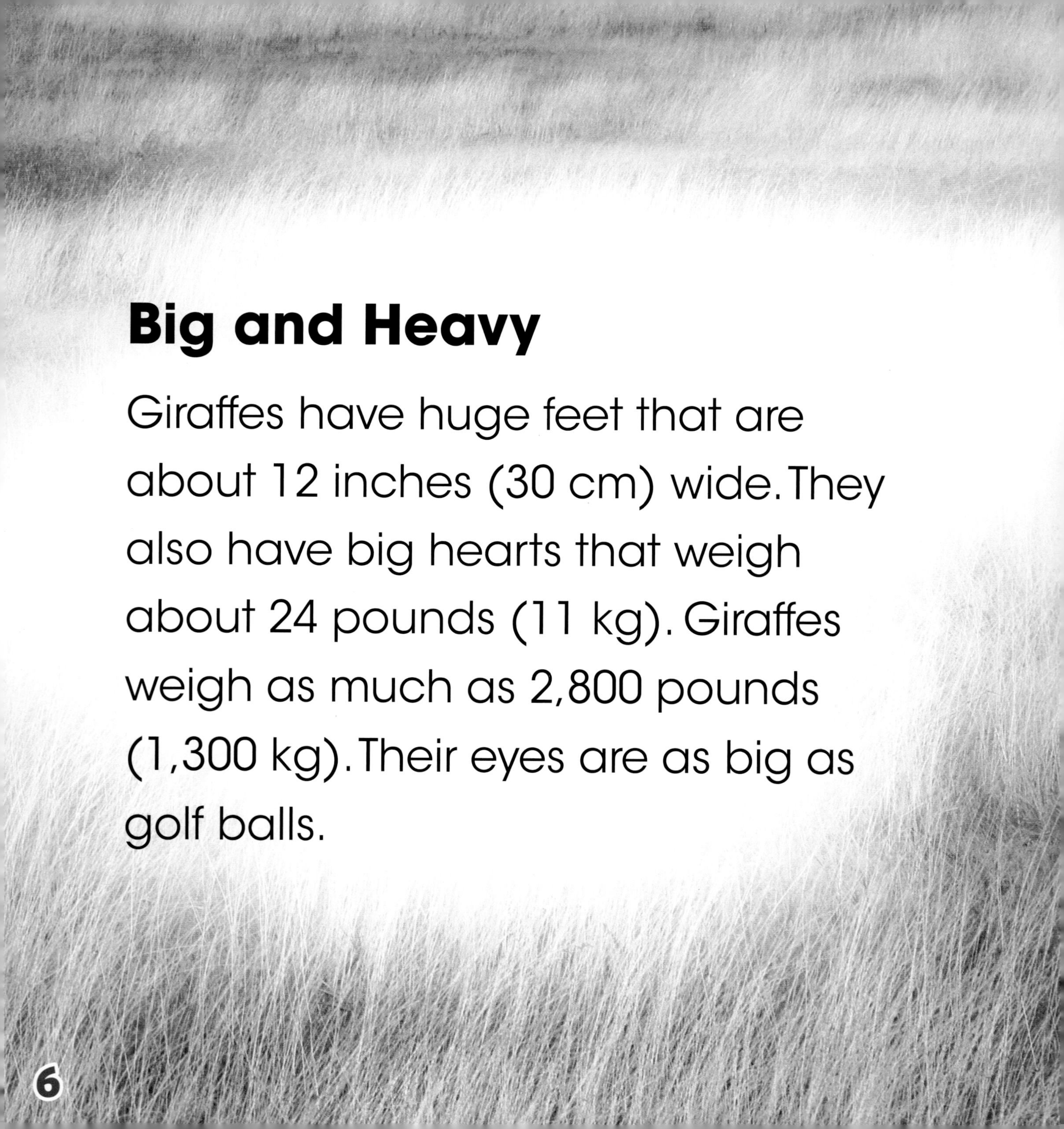

Big and Heavy

Giraffes have huge feet that are about 12 inches (30 cm) wide. They also have big hearts that weigh about 24 pounds (11 kg). Giraffes weigh as much as 2,800 pounds (1,300 kg). Their eyes are as big as golf balls.

Spot the Difference

There are four different species, or kinds, of giraffes. Some are light brown and others are very dark. All have spotted coats. Some people believe that the spotted **pattern** helps giraffes hide among trees. No two giraffes have the same markings.

A Tower of Giraffes

Male, or boy, giraffes are called bulls. Females, or girls, are called cows. Baby giraffes are called calves. Cows aren't as tall as bulls. Giraffes live in groups called towers. A tower has around 15 members.

A Giraffe's Life

Female giraffes give birth standing up. That means the newborns drop about 5 feet (1.5 m) to the ground! Giraffe calves can stand up within an hour of being born. In the wild, giraffes live for about 25 years.

Reaching High

Giraffes are herbivores, or plant eaters. They eat up to 75 pounds (34 kg) of leaves and twigs a day. Giraffes use their long tongues to reach leaves on the tops of trees. A giraffe's tongue is around 20 inches (50 cm) long.

Bending Low

Giraffes drink water only every few days. They get most of their water by eating leaves. Their necks are too short to reach the ground. To drink, they have to bend their knees or spread their front legs wide.

Staying Safe

Male giraffes use their long necks and horns to fight each other over land and **mates**. They use their height and good eyesight to look out for **predators**, such as lions and crocodiles. Giraffes can run up to 35 miles (56 km) per hour.

Giraffes in Trouble

People sometimes kill giraffes for their coats and tails. As human populations rise, giraffe **habitats** are shrinking. People are cutting down the trees that giraffes eat. If they aren't **protected**, giraffes are in danger of dying out in the wild.

Fancy Features

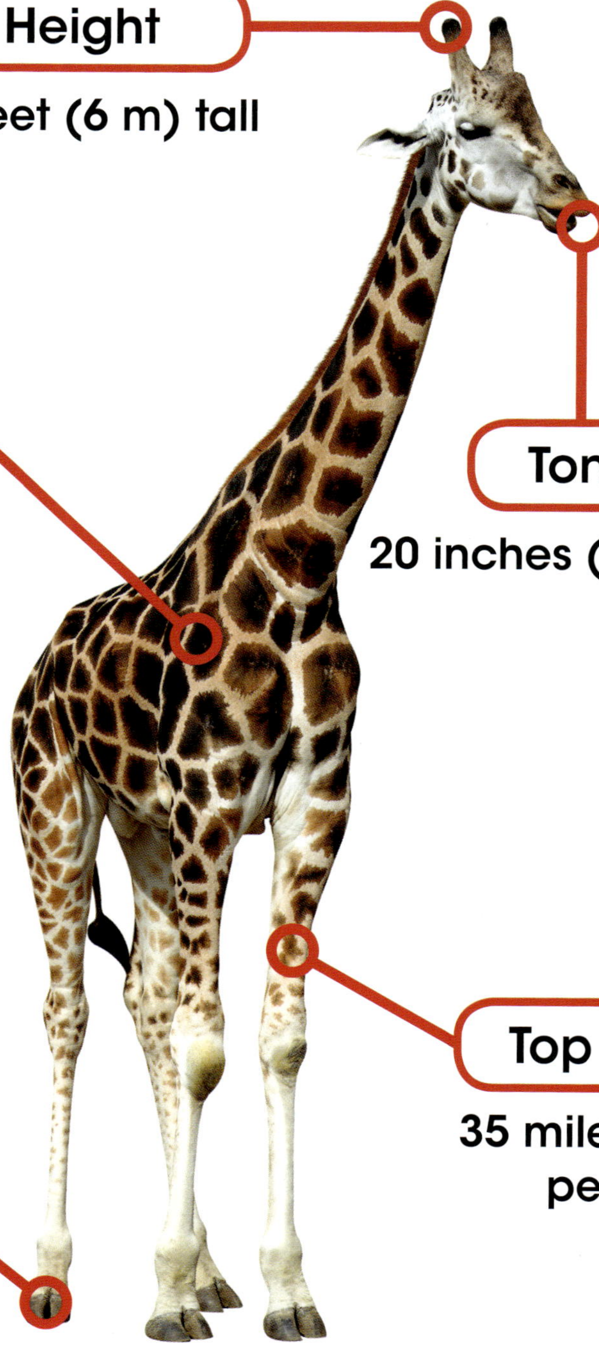

Height

19 feet (6 m) tall

Tongue

20 inches (50 cm) long

Weight

up to 2,800 pounds (1,300 kg)

Top Speed

35 miles (56 km) per hour

Feet

12 inches (30 cm) wide

GLOSSARY

grassland: Land on which grass is the main plant life.

habitat: The natural home for plants, animals, and other living things.

mate: One of two animals that come together to make babies.

pattern: The way colors and shapes appear over and over again on something.

predator: An animal that hunts other animals for food.

protect: To keep safe.

INDEX

WEBSITES

Due to the changing nature of Internet links, PowerKids Press has developed an online list of websites related to the subject of this book. This site is updated regularly. Please use this link to access the list: www.powerkidslinks.com/aotg/giraffes